Understanding Monoclonal Antibodies

Understanding Monoclonal Antibodies

Dr. Sowmya Jess

Copyright © 2020 by Dr. Sowmya Jess.

ISBN:	Softcover	978-1-7960-9752-8
	eBook	978-1-7960-9751-1

All rights reserved. No part of this book may be reproduced or transmitted in any form or by any means, electronic or mechanical, including photocopying, recording, or by any information storage and retrieval system, without permission in writing from the copyright owner.

Any people depicted in stock imagery provided by Getty Images are models, and such images are being used for illustrative purposes only.
Certain stock imagery © Getty Images.

Print information available on the last page.

Rev. date: 06/05/2020

To order additional copies of this book, contact:
Xlibris
1-888-795-4274
www.Xlibris.com
Orders@Xlibris.com

Introduction

In 1975, the breakthrough technology for production of Monoclonal antibodies was published by JK Kohler and Cesar Milstein enabling the immortalization of B cells in order to use them as cell factories for the production of mAbs using the hybridoma technology for which they were awarded the Nobel Prize for Physiology or Medicine. Monoclonal antibodies (mAbs) as opposed to Polyclonal antibodies (pAbs) are specific for a single antigenic epitope, have the exact same sequence of amino acids in their antigen binding regions and are produced by a single clone of B lymphocytes. mAbs and their fragments are a burgeoning group of pharmaceutical molecules. Today the biopharmaceutical market accounts for 40% of the global pharmaceutical market sales. The monoclonal antibody market is supposed to hit $138.6 billion by 2024. Since the licensing of the first monoclonal antibody -Muromonab CD3 for transplant rejection 30yrs ago, the FDA has approved approximately 60 therapeutic monoclonal antibodies (as of 2017) with several in various phases of Clinical trials. Since the year 2000, the therapeutic market for mAbs have grown exponentially. Since 1985, approximately 100 mAbs have been designated as drugs. The WHO which is responsible for therapeutic mAb nomenclature reported in 2017 that over 500 mAb names have been provided. Most are concerned with immunological or oncological targets. Over the years considerable research has been done in the area of oncogenesis which has led to the discovery of a plethora of antigens linked to the development of cancer. Alongside the discovery of these antigens, there has been a surge in the development of monoclonal antibodies intended to target these antigens in the treatment of cancer. Immunotherapy of cancer using naked and conjugated mAbs offer the advantage of markedly decreased adverse effects of targeted therapy sparing normal tissues from serious damage inflicted by conventional chemotherapy.

Production of mAbs

Antibodies can be designed that specifically target a certain antigen such as one found on cancer cells and then many copies of this antibody can be made. The first thing to do is find the right antigen to attack. The target molecule could be a cell – cycle regulatory proteins involved in tumorigenesis. Since growth factors and their receptors frequently play a significant role in the unchecked proliferation of cancerous tissues, several mAbs used in immunotherapy of cancer target these molecules. Some examples include; EGFR: Epidermal growth factor receptor; HER: Human epidermal growth factor receptor; IGFR: Insulin-like growth factor receptor; TGF-βR: Transforming growth factor-beta receptor; VEGFR: Vascular endothelial growth factor receptor; PDGFR: Platelet-derived growth factor receptor; FGFR: Fibroblast growth factor receptor etc.

- EBV transformation of B cells - Several years before the advent of the mouse hybridoma technique, monospecific B Lymphocytes with the desired antigenic specificity were transformed by EBV (Epstein Barr Virus) into immortalized cell lines which would continue to produce the desired antibody. This technique (mutagenic transformation of monoclonal B cells of desired specificity by an oncogenic virus) is rarely used now. [Epstein Barr virus is the causative agent of Infectious mononucleosis (IMN) and is responsible for the malignant transformation of B cells in Burkitt's lymphoma.]
- Mouse Hybridoma technique- mAbs against a variety of antigens can be produced by immunization of mice. Mouse is immunized with the antigen (against which a therapeutic mAb is being

developed). The animal is euthanized, its spleen harvested and the B cells producing the desired antibody are isolated. This B cell is then fused with a mouse myeloma cell. This somatic cell hybridization of a normal B lymphocyte with a malignant cell (the myeloma cell) results in cell hybrids (tetraploid hybridomas) that have the preserved specific antibody secretory function of the B cell and the growth properties of the malignant cell. These hybridomas are then grown in a medium known as the HAT medium which kills only the unfused myeloma cells (which might otherwise outgrow the weaker tetraploid hybridoma cells). The unfused B cells have limited power of division and will die off naturally in culture. The HAT medium is thus selective for the mAb secreting hybridoma cell.

[Mouse myelomas or plasmacytomas are produced by injecting mouse with a tumorogenic agent.

The mouse plasmacytoma/myeloma cell bears a designation consisting of an alphabetical prefix usually derived from the tumorogenic agent or the last initial of the investigator who induced the tumor sometimes combined with PC (for plasmacytoma).

E.g. TEPC mouse myeloma cell line in which the tumorogenic agent is TE (Tetramethylpentadecane); ABPC where the tumorogenic agent is Abelso virus (AB).]

The antibody secreting hybridomas are then injected into the abdominal cavity of a mouse and the mAbs are harvested from the ascitic fluid. The ascites method of mAb production causes significant pain and discomfort to the animals and has been banned in several countries in favor of in – vitro methods.

- A variation on the conventional mouse hybridoma technique is to replace the Fc region (constant region) of the immunoglobulin gene in the mouse with human Fc gene so that the resultant antibody (a chimeric human-mouse fusion protein) would be less immunogenic. This is one of the techniques for generating hybrid immunoglobulins.

-Heterohybridoma using heteromyeloma fusion partners- It is possible to fuse an antigen- specific B cell from one animal with the myeloma cell of another animal. E.g. fusion of monoclonal antibody producing clone of porcine B cells with murine myeloma cells. The fact that a hybridoma can be produced by fusion of cells from two different species is interesting from an academic point of view but despite digging through a lot of journals on PubMed, I haven't been able to come across any human therapeutic applications for this discovery, although the ability to fuse antigen – specific human B cell with myeloma cells from another species would have therapeutic potential because of the ability to generate fully human antibodies.

- Human Hybridomas – Fusion of human B cells producing the desired Ab with human myeloma cells can be achieved. Although this has the advantage of producing fully human mAbs, the human hybrids grew slower and had a higher degree of chromosomal instability compared to murine hybridomas. Thus, only a small proportion of human hybridomas are stable and produce the antibody of the desired specificity. Some interesting human hybridomas have been produced: -
- Human T cell – B cell hybridomas containing 92 chromosomes have been formed by fusion of human T cells with EBV transformed B cells. These hybridoma cells variably expressed surface IgM BCR (B cell receptor) and TCR (T cell receptor) and could be stimulated to produce IgM antibodies.
- Patients with PN- MGUS (Peripheral neuropathy monoclonal gammopathy of unknown significance), also known as Anti-MAG (Myelin associated glycoprotein) peripheral neuropathy, secrete

autoantibodies that attack peripheral myelin. Immunizing mice with these IgM anti-MAG antibody induces the secretion of mouse anti-idiotype antibody known as A8F2 against this IgM anti- MAG Ab (When an antibody binds to the idiotype of another antibody, it is referred to as an anti-idiotype Ab. The variable part of an antibody including its unique antigen binding site is known as the idiotype). Thus, A8F2 is a monoclonal anti-idiotype antibody that recognizes and binds to the MAG binding site of the patient's Anti-MAG Ab. It shows high affinity binding to the antigen- binding region of the patient's anti-MAG auto antibody and prevents its binding to the MAG antigen on the peripheral myelin sheath. Mouse B cells producing A8F2 are fused with mouse myeloma cells and the resultant murine hybridoma cells secrete the anti- idiotype Ab-A8F2. (Although included under human hybridomas, please note that this is a murine hybridoma)

- A human hybridoma was produced by fusing a human B cell from a patient suffering from MM (Multiple myeloma) with peripheral B lymphocytes from a patient with SSPE (Subacute Sclerosing Pan Encephalitis – a measles prion infection). These hybrids were found to secrete human IgM specific for the measles virus capsid.
- Human-human hybridomas were obtained by fusing lymph node lymphocytes with human malignant B cells. Lymphocytes were obtained from lymph nodes draining the primary breast tumor and were fused with human B-lymphoblastoid cells. The resultant hybridomas produced IgG/IgM antibodies against autologous malignant breast tissue. One of these mAbs (MAC 40/43) reacted specifically with a glycoprotein associated with breast and colon carcinomas and other neoplasms of epithelial origin. They did not react with normal breast tissue or any other tissue.

The mAbs secreted by murine hybridomas have their limitations in human therapy owing to their immunogenicity. Humanization of the murine Ab for human therapeutic use would incur extra cost. Thus, B cell hybrids that secrete human mAbs are of considerable interest. Although human hybridomas can secrete fully human mAbs, the process is cumbersome and the yield low. The Phage Display technology (APD – Antibody Phage Display) overcomes some of these limitations.

- Phage Display technology -This is used for the production of antibody fragments (such as variable region fragments) rather than for the production of whole antibody. Efforts over nearly two decades have indicated that antibody fragments (Fab or Fv) can only rarely be prepared by proteolytic dissection of whole IgG or IgA antibodies. The more expedient method for producing antibody fragments is the Phage Display Technology. The animal (mouse) is challenged with the antigen against which the mAb is being produced. mRNA corresponding to the variable chain (antigen binding) region of the antibody is isolated from its splenic lymphocytes. (mRNA can also be isolated from peripheral human B lymphocytes for the production of fully human Ab fragments through phage display. However, hyperimmunization of human beings to generate antibody producing B cells of desired specificity is considered unethical). Using the enzyme reverse transcriptase c-DNA (complementary DNA) is produced on the m-RNA templates. The DNA is then amplified using PCR (Polymerase Chain Reaction- the segment of DNA is put in a solution containing DNA polymerase, primers and nucleotides. The process can produce billions of copies of the gene in a couple of hours and is faster than cloning in E. coli. The c-DNA is then incorporated into the genome of filamentous bacteriophages next to the pIII or pVIII genes encoding for the minor or major coat protein respectively. Now the antigen- binding Ab fragment (corresponding to the c-DNA incorporated into the viral genome) is displayed on the viral capsid as part of its protein coat. DNA encoding for millions of ligands (including ab fragments and other peptides can be introduced into the phage genome this way. By modifying the c-DNA generated from the m-RNA corresponding to the VH and VL (Variable region, Heavy and Light chains) immunoglobulin regions of splenic or peripheral lymphocytes and incorporating them into the phage genome, you can build an APD (Ab Phage Display) library exhibiting a vast repertoire of Ab fragments with different antigenic specificities and with improved biophysical characteristics such as higher affinity, greater stability and lesser immunogenicity. For e.g. there was a paper published in which a two amino acid modification in the CDR region (complementarity – determining region) of the variable region of the antibody greatly reduced immunogenicity while retaining full biologic activity. In order to select the phage displaying the Fv fragment with the desired specificity (i.e. binds to the target antigen), a process called Panning is used. Microtiter wells are coated with the desired antigen, phages are incubated with the antigen and any unbound phage is washed away. The bound phages are the ones displaying the Fv fragment with the desired specificity and these are allowed to infect E. coli. The phage genome integrates into the E. coli genome (a process called Transduction) which then serves as a Cloning and Expression system if the phage follows a lysogenic cycle. The expression system/ expression host secretes the antibody fragment which can be harvested. (This process is similar to the production of recombinant human insulin). The first clinically approved therapeutic antibody obtained with the help of phage display was adalimumab (marketed as Humira®), which neutralizes tumor necrosis factor and is mainly used against rheumatoid arthritis. Today, it is the best-selling biological.

[When a bacteriophage is used to transfer a gene into a prokaryotic cell/expression system like E. coli, the process is called transduction. A similar process in a eukaryotic cell is called transfection. The commonly used expression vectors are   E. coli (prokaryotic), Saccharomyces cerevisiae (eukaryotic) and Chinese Hamster Ovary (CHO) cells (Mammalian cell line). Microorganisms can be used as biopharmaceutical production factories. E. coli are the expression host of choice in the production of antibody fragments such as variable region fragments. Antibody fragments like Fab (the antigen binding fragment) still exhibit antigen binding properties and have the advantage of being able to be produced in microbial organisms which are easy to manipulate and cultivate. But in some cases, it may be necessary to produce the whole Ab including the Fc portion. This is because the Fc region is important in antibody effector functions such as CDC (Complement Dependent Cytotoxicity) and ADCC (Antibody Dependent Cellular Cytotoxicity) and phagocytosis. Multiple polypeptide proteins such as antibodies are difficult to express in prokaryotic systems such as E. coli due to the complexity of protein folding plus secretion. Traditionally therefore, full length antibodies are expressed in mammalian cell culture like CHO or HEK (Human Embryonic Kidney) Cell lines. (Mammalian cell lines allow human like N-glycosylation which is needed for the production and secretion of full-length antibodies). The downside however is that recombinant DNA expression in mammalian cell lines is a complicated technology, is costly and has a low yield compared to prokaryotic expression vectors.

[A paper has however been published which demonstrated the successful expression of full-length IgG in an E. coli strain HB2151!]

- Production of chimeric and humanized antibodies by grafting of specific antibody segments onto a scaffold – A chimeric antibody is formed by replacing the constant region of the mouse antibody with a human Fc portion reducing immunogenicity. So in a chimeric antibody, the Fc portion is human while the Fv portion is murine.
- Humanization involves going a step further in which only the CDR portions of the murine variable region (not the entire variable region) are grafted on to a human antibody scaffold. The CDR portions of the murine Ab will have to be retained as these determine the specific binding ability of the Ab to the target protein i.e. it is essential to the integrity of the antigen combining site. However even the xenogeneic CDR regions of the humanized antibody may evoke an anti – idiotypic response in patients. To minimize this a new approach to antibody humanization was developed called SDR (specificity determining residues) grafting. SDRs are regions within the CDR which are most crucial to the antibody-ligand interaction. SDRs are determined through database analysis of three-dimensional structures of antigen-antibody complexes. In this type of humanized antibody (which is a step closer to fully human), the xenogeneic regions are confined to the relatively small SDR segments, further reducing the immunogenicity in humans.

- Production in transgenic animals- Fully human antibodies can be developed in transgenic mice which have been genetically engineered with the human immunoglobulin gene which replaces the murine immunoglobulin gene. DNA microinjection methods are used to create transgenic animals. In DNA microinjection also called Pro-nuclear injection a very fine glass pipette is used to manually inject DNA into the zygote early after fertilization when there are two pronuclei. After fusion of the two pronuclei to form the zygote, the injected DNA may or may not be taken up. These transgenic mice with the human immunoglobulin gene, upon antigenic stimulation produce fully human immunoglobulins thereby reducing the potential for allergic reactions due to murine residues.

- Production in SCID mice – Mice homozygous for the SCID (Severe Combined Immunodeficiency Disease) are severely deficient in functional B and T cells. These mice can be grafted with human peripheral B lymphocytes at a young age (when their immune system is immature reducing the probability of rejection of human tissue) i.e. they can be populated with an artificial human immune system. Such mice when hyperimmunized with the antigen produce B lymphocytes making human antibodies which can be fused to a human myeloma cell to generate a hybridoma that produces human antibodies.

- Chemical synthesis of DNA- In this process a DNA strand is synthesized from free nucleotides without the aid of a template strand or DNA polymerase enzyme (both of which are essential for DNA synthesis through the PCR method) using a machine known as the Gene Machine. Prior to this the nucleotide sequence of the gene of interest is deduced by a DNA sequencer which is an automated machine used to analyze and determine the sequence of bases in a sample of DNA. The Gene Machine also called an automated polynucleotide synthesizer can synthesize a gene (e.g. an immunoglobulin gene) from a pre-determined sequence rapidly and in high amount in a few days. There are four separate reservoirs for each of the nucleotides – A, T, C and G. The synthesis of a strand of DNA is under the control of a microprocessor. The desired sequence is entered on a keyboard and the microprocessor automatically opens the valve of the desired nucleotide which permits its entry into the synthesizer column where the actual strand is synthesized at the rate of two nucleotides per hour. [By feeding the instructions for the human insulin gene into the gene machine, this gene has been synthesized, and incorporated into E. coli plasmid for the production of human insulin].

The DNA sequence (gene) corresponding to any protein can be deduced by isolating the corresponding m-RNA→ c-DNA→ dsDNA. The dsDNA is then sequenced using a DNA sequencer. Some Sequencers can sequence single stranded DNA. GenBank and ENA (European Nucleotide Archive) are repositories of publicly available nucleotide sequences and are readily accessible online. GenBank is the NIH (National Institute of Health) genetic sequence database. GenBank and ENA are part of the International Nucleotide Sequence Database Collaboration. It receives contributions from laboratories around the world. Anyone interested in the DNA sequence of a protein can just go to this library and pull up the genetic sequence of that protein. This information would be useful in the chemical synthesis of

DNA. As of now 32 million entries are available at GenBank. {https://www.ncbi.nlm.nih.gov/guide/howto/find-transcript-gene/}. Therefore, if you want to synthesize a gene whose sequence has already been decoded, you can access its sequence from GenBank, and then use the microprocessor to input the sequence into the gene machine to artificially synthesize a gene, for applications like incorporation into a plasmid.

Staggering advances have been made in the area of DNA sequencing technology and is giving us greater insight into the genetic basis of a variety of diseases. It is estimated that soon the cost for sequencing someone's genome can be about $100 and be completed within about 1 hour! Contrast this to the Human Genome Project which was launched in 1990 with multiple countries participating in sequencing the first human genome which took 15 years and cost about $3 billion.

Nomenclature of monoclonal antibodies

Active pharmaceutical substances/ingredients (APIs) including biologics like monoclonal antibodies require an International Nonproprietary name (INN) assigned by the WHO for the manufacturer to obtain marketing authorization. INNs for pharmaceuticals are unique, generic, unambiguous names for identification by practitioners worldwide. The USAN (United States Adopted Names) Council is also instrumental in assigning antibody names. The Antibody society (Tabs), an international, non-profit association involved in antibody related research and development also provides their input regarding mAb nomenclature. The nomenclature for mAbs has been undergoing continuous revisions making it an arduous task to keep pace with the rapid revisions in criteria.

Monoclonal antibody nomenclature is different from that of other pharmaceuticals. I shall briefly go over the basics of monoclonal antibody nomenclature. mAb names are composed of morphemes or infixes or sub stems. Here I shall use the word substems.

Until 2017, the INN of mAbs was assembled from a prefix that could be freely chosen, followed by a substem A indicating the target E.g. "tu" for tumor; "ci" for circulatory system; "lim" for immune system etc. (Regarding this substem there is a difference between the old and new naming conventions which I shall not addressing here). Substem B indicates the source ;-

-o- Mouse (-omab) (murine antibody)

-xi- Chimeric (-ximab) (murine Fv portion grafted on to human Fc portion)

-zu- Humanized (-zumab) (only the hypervariable or CDR regions are murine)

-xizu- Chimeric Humanized (-xizumab) (combination of both chimeric and humanized)

-u- Human (-umab) (fully human)

-a- Rat

-e- Hamster

-i- Primate

Examples: -

-Trastuzumab, Tras-tu-zu-mab (Tras is the prefix, "tu" is the target substem indicating that it targets a tumor in this case Ca Breast, "zu" is the source substem indicating that it is humanized and "mab" is the stem indicating that it is a monoclonal antibody.

-Abciximab; Ab- prefix, ci- acts on circulatory system, xi- chimeric, mab- monoclonal

-Rozrolimupab; Rozro- prefix, lim- acts on immune system, u- fully human, pab- polyclonal antibody

- Muromonab CD3, the first mAb to be licensed by the FDA was licensed 30 yrs back before the naming convention for mAbs was introduced and so ends with a "nab" instead of the "mab" which is now the standard stem for mAb nomenclature.

The aforementioned nomenclature was used until 2017 after which there was a unanimous decision to discard the "Source substem or substem B" containing species information. In June 2017, WHO announced that the use of the source infix/substem will be discontiued for new antibody INNs effective immediately.

Mouse (murine) IgG antibody (Substem B would be -o-)

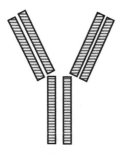

Murine residues are shown as stripes and human sequences are shown in grey.

Chimeric antibody with mouse variable region grafted onto human antibody scaffold (both the Fc portion and the C_{H1} and C_L regions are human) (Substem B would be -xi-)

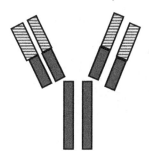

Humanized antibody where only the CDR (complementary determining regions) are murine (Substem B would be -zu-)

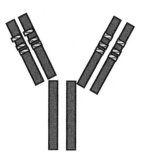

Fully human antibodies with no murine residues (Substem B would be -u-)

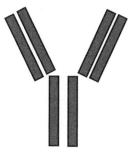

Combination of both humanized and chimeric antibodies (Substem B would be -xizu-)

Whole antibodies and antibody fragments

Monoclonal antibodies can be whole monoclonal antibody gererally of the IgG or Ig M type or Anibody fragments. Whole antibodies have the advantage of retaining Fc mediated antibody effector functions as mentioned previously. They have the disadvantage that mammalian cell lines are required for production making the process cumbersome. Antibody fragments retain the antigen-binding capacity (although they lack the Fc mediated effector functions) and can be easily produced using phage display technology.

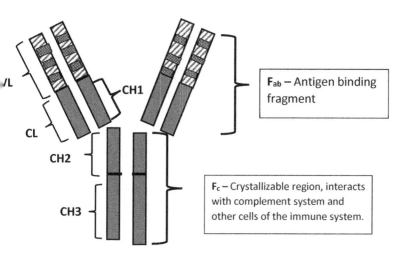

F_{ab} – Antigen binding fragment

F_c – Crystallizable region, interacts with complement system and other cells of the immune system.

Structure of Whole IgG antibody

- The heavy and light chains of the Fab fragment are linked by disulfide bonds.
- There are two flexible amino acid chains that link the CH1 and CH2 regions of the heavy chains, that are interconnected by disulfide bonds.
- This forms the hinge region of the amino acid molecule.

- ■ – Constant regions, heavy and light chains
- ▨ – Variable regions, heavy and light chains
- ▦ – CDR regions

Antibody fragments

Fab – It stands for antigen-binding fragment- one arm. It comprises of VL, CL, VH and CH1.

Solid grey represents Light chain and Pattern fill represents heavy chain Fab regions

Fab' – Same as Fab except that it includes the hinge region

(Fab')$_2$ – Two Fab fragments linked together by the hinge region.

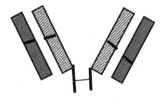

scFv – Consists of VL and VH of one chain linked together.

Di – scFv – Consists of scFv of both chains linked together. Therefore, consists of two VH and two VL regions. They are also called *Diabodies*. Multivalent antibody fragments can be produced to increase binding affinity by linking together scFv fragments. *Triabodies* are produced by linking together three scFv fragments. *Tetrabodies* are produced by linking together four scFv fragments.

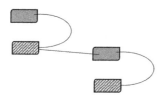

sdAb – Also known as VHH or nanobody, is composed of the Fab fragment of camelid antibody. The Fab region of camelid antibody consists solely of Heavy chain variable region (termed VHH) without the Heavy chain constant region (CH1) or the light chain regions as in humans. Thus, the nanobody technology was originally developed following the discovery that certain animals like camels have fully functional antibodies consisting only of heavy chains. Nanobodies are a new generation of novel single domain antibody fragments.

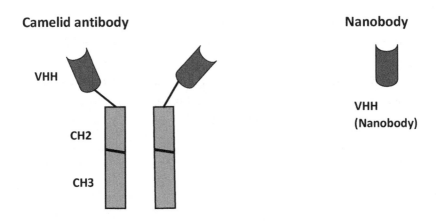

Nanobody molecules are small (12-15 KDa), robust and their sequence homology is comparable to humanized mAbs. Their small size and the fact that they can be readily linked to toxins, drugs, radionuclides, photosensitizers etc. are properties that make them particularly suitable for specific and efficient targeting of tumors in vivo.

Ablynx is a subsidiary of biopharmaceutical company Sanofi engaged in the discovery and development of nanobodies.

BiTE – Bi-specific T-cell engager (These bispecific molecules are created by linking the targeting regions of two antibodies. One arm of the molecule is engineered to bind with a protein found on the surface of cytotoxic T cells, and the other arm is designed to bind to a specific protein found primarily on tumor cells. When both targets are engaged, the BiTE® molecule forms a bridge between the cytotoxic T cell and the tumor cell, which enables the T cell to recognize the tumor cell and fight it through an infusion of toxic molecules. The tumor-binding arm of the molecule can be altered to create different BiTE® antibody constructs that target different types of cancer.

Trifunctional Bispecific antibody – It is a Bifunctional antibody in that it has two arms that can bind to two different receptors (one on the cytotoxic T-cell and the other to a protein on the tumor cell). In addition, it has an Fc portion which can bind to receptors on Macrophages, NK cells and dendritic cells. Because of the presence of the Fc region, these molecules retain antibody mediated effector functions such as complement fixation (complement-dependent cytotoxicity, CDC) and Antibody-dependent cellular phagocytosis (ADCP).

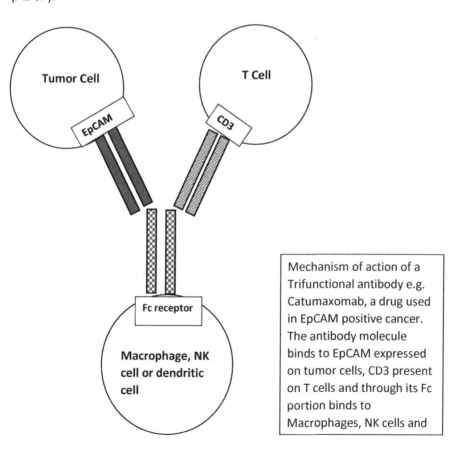

Mechanism of action of a Trifunctional antibody e.g. Catumaxomab, a drug used in EpCAM positive cancer. The antibody molecule binds to EpCAM expressed on tumor cells, CD3 present on T cells and through its Fc portion binds to Macrophages, NK cells and

Naked and Conjugated monoclonal antibodies

Naked mAbs are antibodies that work by themselves. There is no drug or radioactive material attached to them. These are the most common type of mAbs used to treat to cancer.

Conjugated monoclonal antibodies – mAbs joined to a chemotherapy drug (chemolabeled mAbs) or to a radioactive particle (radiolabeled mAbs) are called conjugated mAbs. The mAb is used as a homing device to take one of these substances directly to the cancer cells. This lessens damage to normal cells in other parts of the body. They are also called tagged, labelled or loaded antibodies.

Mechanism of Action and some Therapeutic indications

mAbs have found use both in the field of Immunodiagnostics and Immunotherapy. Common target molecules for mAbs are the CD20 Ag on B-cells, CTLA4 on T-cells; Growth factor receptors such as HER2 (Human Epidermal Growth Factor Receptor2), EGFR (Epidermal Growth Factor Receptor); VEGFR (Vascular Endothelial Growth Factor receptor); EpCAM (Epithelial cell adhesion molecule which is a transmembrane glycoprotein involved in cell signaling and proliferation); Ligands such as TNF-alpha (Tumor necrosis factor α), VEGF (Vascular endothelial growth factor). New targets for mAbs are continuously being discovered as researchers are finding more and more molecules linked to cancer, autoimmune and other diseases.

- mAbs to hGH are used in the RIA (Radioimmunoassay)of hGH (human growth hormone) levels in serum. Hybridomas (which produced antibodies against hGH) were generated after immunization of mice with hGH,. mAbs derived from these,

recognized three different epitopes on the hGH molecule. One of these showed low cross- reactivity with the related hPL (human placental lactogen). This mAb has been used to establish RIA for serum hGH measurements and show high sensitivity (due to the high affinity) and high specificity (due to low cross- reactivity).
- MG7 mAb, a monoclonal antibody directed against MG7Ag (a gastric cancer associated antigen distinct from CEA) could be used as a screening tool for gastric cancer. The antibody is produced by immunization of mice with MG7Ag. Interestingly, it was found that when phages displaying certain nanopeptide fragments were incubated in wells coated with MG7 mAb, some of them bound to the antibody. These nanopeptides were therefore mimotopes of the MG7Ag. When these MG7 mimitope nanopeptides were subsequently injected separately into mice, it was discovered that some of them induced the production of an MG7 antibody thereby making them immunogenic mimitopes of the MG7Ag.
- In mice infected with Influenza virus H5N1, nanobodies directed against hemagglutinin (hemagglutinin and neuraminidase are the two glycoproteins on the influenza virus membrane) were effective in controlling the infection.
- Despite the fact that HBV vaccination is now widely available, HBV infection is still responsible for a considerable number of cases of chronic hepatitis leading to cirrhosis and in some cases hepatocellular carcinoma. Although antiviral drugs like Entecavir can suppress viral replication, they cannot eradicate the infection. HBV specific monoclonal antibodies which would bind to the envelope proteins of the virus are being explored as a possible adjunct to HBV anti-viral therapy.
- Photothermal therapy - Nanobodies binding to HER2 antigen which is overexpressed in breast and ovarian cancer cells were

conjugated to gold nanoparticles and the tumor was destroyed photothermally using laser since gold nanoparticles absorb light energy and create heat thereby destroying the cancer cells. The technique has also been used against microorganisms.
- Nanobodies might cross the BBB and permeate into solid CNS tumors more easily than whole antibodies or larger antibody fragments which would allow the targeted delivery of drugs against brain cancers like Astrocytomas.
- RIT - Ibritumomab tiuxetan (Zevalin) is a radiolabeled mAb used for the treatment of NHL. The mAb Ibritumomab is attached to the radioisotope yttrium-90 by the chelator Tiuxetan. Ibritumomab attaches to the CD20 antigen found on the surface of B cells allowing radiation from the attached isotope to kill it. This type of cancer therapy is known as Radioimmunotherapy (RIT).
- ADC - Trastuzumab emtansine also known as Ado-trastuzumab emtansine (Kadcyla) is an Antibody Drug Conjugate (ADC). The antibody, Trastuzumab targets the HER2 protein overexpressed on certain breast cancers and delivers the potent anti-mitotic agent DM1. These antibodies are also known as chemolabeled antibodies.
- In ADEPT (Antibody-directed enzyme prodrug therapy), an enzyme conjugated to an anti-tumor antibody or antibody fragment is given IV and localizes in the tumor. A prodrug is then given which is converted to a cytotoxic drug selectively in the tumor. In clinical trials, patients with colorectal cancer received ADEPT therapy with A5B7 which is an F(ab')$_2$ antibody fragment against CEA (Carcinoembryonic antigen) conjugated with the enzyme CPG_2 (Carboxypeptidase G_2). The patient was then injected with the pro-drug – a benzoic acid mustard glutamate,

which was converted into the active metabolite and reached cytotoxic levels within the tumor while remaining at much lower concentrations in the plasma thus sparing normal tissues.
- Immunoliposomes provide a complementary and advantageous drug delivery strategy to ADCs (Antibody Drug Conjugates). They have a high carrying capacity (20,000 – 150,000 drug molecules per liposome. Liposomes are vesicular particles composed of a lipid bilayer encapsulating a hydrophilic drug in the interior or lipophilic drug into the lipid bilayer. They can be used as drug delivery systems for cancer drugs – E.g. Anti – HER2 Immunoliposomes.
- Check-point therapy – A check-point inhibitor is a drug that blocks certain proteins that hamper the immune system (e.g. T cells) from attacking the tumor cells. E.g. the protein PD-1 on T-cells binds to the protein PD-L1 on tumor cells acting as immune check-point proteins. They help tumors evade immunosurveillance. Anti-PD1 and Anti-PD-L1 antibodies/antibody fragments thus act as immune checkpoint inhibitors. [PD-1 is a checkpoint protein on T-cells. When PD-1 binds to a protein called PD-L1, which is present on normal and cancer cells, the T-cell is turned off. Some cancer cells produce large amounts of PD-L1 which bind to PD-1, thereby turning off the T cell and evading immunosurveillance. mAbs called check point inhibitors that target either PD-1 or PD-L1 can block this binding and boost the immune response against cancer cells.
- Muromonab CD3 (OKT3), a murine mAb was the first FDA approved therapeutic mAb. It was approved for the prevention of transplant rejection in 1986. However, because of serious adverse effects like severe cytokine release, and high degree of immunogenicity of mouse antibody in humans apart from severe

immunosuppression, and owing to the availability of better alternative therapies, it was withdrawn from the market in 2010.
- Later, the first chimeric mAb, Rituximab was approved for the treatment of low-grade B-cell lymphoma.
- Ixekizumab, a form of humanized mAb which neutralizes IL-17A is approved for the treatment of moderate to severe plaque psoriasis.
- Erenumab, is a fully humanized mAb used for the treatment of migraine. It acts by blocking the CGRP (Calcitonin Gene Related Peptide) receptor.
- Etanercept (Enbrel) is a fusion protein combining the ligand-binding portion of Human TNF-α receptor (TNFR) with the Fc portion of IgG. The molecule thus binds to circulating TNF-α and prevents it from binding to its receptor on cells. It is effective in the treatment of Rheumatoid arthritis.
- Caplacizumab, a nanobody targeting von Willebrand factor is in clinical trials for the prevention of thrombosis in patients with a/c coronary syndrome.
- Bevacizumab (Avastin) is a mAb that targets VEGF thus hampering neoangiogenesis in tumor growth. Intravitreal (ITV) injection of Bevacizumab is used to treat neovascular AMD (age related macular degeneration) in which VEGF causes excess proliferation of leaky blood vessels growing from the choroidal region into the retina causing loss of vision.
- Cetuximab (Erbitux) is an mAb that targets EGFR.
- Natalizumab is a humanized mAb, the first in a class of DMTs (Disease Modifying Therapies) approved for the treatment of MS, known as selective adhesion molecule (SAM) inhibitors. Endothelial cells in the lumen of blood vessels express Vascular Adhesion Molecule-1 (VCAM-1) at sites of active MS lesions. T

cells express an adhesion molecule known as $\alpha_4\beta_1$-integrin. T cells cross the BBB, enter the CNS lesion where interaction occurs between VCAM-1 and $\alpha_4\beta_1$-integrin. By binding to $\alpha_4\beta_1$-integrin and blocking this interaction, Natalizumab reduces inflammation and formation of MS lesions.
- The most recent FDA approved mAb is SARCLISA (Isatuximab-irfc) for the treatment of adult patients with Multiple Myeloma, approved on March 2nd 2020.
- REGNEB3 is a monoclonal antibody that is pending approval for Ebola virus infection.
- With nanobodies showing efficacy against H5N1 in mice and REGEB3 pending approval for Ebola virus infection, developing monoclonal antibody/antibody fragment against the S1 domain of the spike protein of SARS-Cov-2 (Covid-19) is within the realm of possibility.
- **Could mAbs ultimately prove to be the therapeutic intervention of choice for the treatment of Covid -19?**

SARS-Cov-1 and SARS-Cov-2 are closely related betacoronaviruses similar to the Sarbecoviruses isolated from bats. They express a large (approximately 140 kDa) glycoprotein termed Spike protein (S Protein) which mediates binding to host cells via interaction with ACE2 receptor. The S protein is highly immunogenic with the RBD (Receptor Binding Domain of the S protein which is responsible for most of the interaction with antibodies). From studies of convalescent serum, it has been documented that antibodies mounted upon infection target the full-length S-protein as well as the RBD, the latter being the major target. Patients recovering from Covid-19 disease can be screened for strong antibody response using ELISA. The titer of anti- RBD antibodies correlate better with virus neutralization.

This is from a paper published in MedRxiv
https://www.medrxiv.org/content/10.1101/2020.03.17.20037713v1.full.pdf

They generated two different versions of the Spike protein or S-protein. The first one expresses the full length S-protein and the other the much smaller Receptor Binding Domain (RBD){similar to full length IgG and Fab fragments}. The sequence used was based on the genomic sequence of the first virus isolate, Wuhan-Hu-1, which was released on Jan 10th 2020. The full length spike protein sequence was modified (mutations were induced) to generate recombinant versions of the S protein which had more stability (a property that would be useful for vaccine development).

There are two expression vectors that have been used for the production of S-protein and RBD – The pCAGGS vector for expression in mammalian cells and the Baculovirus vector for expression in insect cells. {An expression vector or expression construct is usually a plasmid or virus designed for gene expression in cells. The gene of interest is first introduced into the vector and the vector is then introduced into the target cell where it integrates into the host genome. The gene is inserted into the vector alongside certain sequences known as promotor and enhancer sequences which lead to increased transcription of the gene of interest.}. The pCAGGS vector is a dsDNA plasmid for expression in mammalian expression systems. Baculoviruses are viruses that infect certain insects like moth and butterflies and the are used as a vector for expression in insect cell lines. For expression in mammalian cells, the sequences were codon optimized [codon optimization is a process used to improve gene expression and translational efficiency of a gene of interest. The mammalian cell codon optimized nucleotide sequence coding for the

Spike protein of SARS-Cov-2 was synthesized commercially (GenBank:908947.3)].

In mammalian cell lines, the RBD domain gene gave outstanding results with remarkably high yield of the protein. The expression of the full-length S-protein gave much lower yield in both mammalian and insect cells.

These proteins have two potential applications. The first is in the area of protein vaccine development (protein vaccines are purified protein antigens of microorganisms, whereas mRNA and DNA vaccines are nucleotide sequences coding for the protein antigen of interest and are introduced through a vector). These vaccines lead to the development of active immunity (as opposed to passive immunity from convalescent sera). Several different types of mRNA vaccines are under trial at present.

The second application is in the area of monoclonal antibody development. Mice can be hyperimmunized with S or RBD proteins and B cells that produce antibodies specific to these immunogenic epitopes of the SARS-Cov-2 virus can be used to generate mAbs through the hybridoma technique as previously described. Monoclonal antibody fragments against S and RBD proteins can be produced by isolation of mRNA from the antigen specific B cells, the complementary DNA synthesized and incorporated into filamentous phages or other expression vectors as described previously. Currently, there are about 10 monoclonal antibodies being evaluated in clinical trials for the treatment of Covid-19. These are all monoclonal antibodies that have been approved for clinical use for a variety of diseases like RA (Rheumatoid Arthritis), PNH (Paroxysmal Nocturnal Hemoglobinuria) etc. and are now being explored for their potential for treating SARS-Cov-2. They target a number of antigens like, IL-6, IL-6R, GMCSF

(Granulocyte Monocytes Colony Stimulating Factor), C5a (a component of the complement system, CCR5 (a chemokine receptor found on the surface of monocytes and lymphocytes) etc. all of which have been implicated in the hyperinflammation associated with Covid-19, an almost invariable predictor of mortality.

After screening 300 antibodies (based on their ability to bind to the S-protein of the virus, Celtrion (a biologics company), identified 38 neutralizing antibodies (which inhibited the virus in culture) of which 14 were culled out for highest potency against SARS-Cov-2.

The Antibody Society is maintaining a list of 60 recombinant biologics including mAbs that target SARS-Cov-2. For further details and a complete list please check out https://www.antibodysociety.org/covid-19/ (Possible Biologic Candidates for EUAs- Emergency use authorizations) in 2020.

Here I will briefly mention 3 of them, Tocilizumab, Leronlimab and STI-1499.

- Tocilizumab, a monoclonal antibody which acts as an IL-6 receptor antagonist and is used as a DMARD (Disease modifying anti rheumatic drugs) in the treatment of RA (Rheumatoid arthritis) is now in FDA approved Phase III clinical trials for the treatment of severe COVID- 19 pneumonia. It is already being used in the US for severely ill patients through the FDA's Compassionate Use Program also known as Expanded Access (this is a pathway for using a drug outside of clinical trials during the pre-approval phase). It binds to both soluble and membrane bound IL-6 receptors (sIL-6R and mIL-6R). The importance of cytokine antagonists in COVID-19 stems from the fact that mortality in most cases is due to virally driven hyperinflammation characterized by fulminant and fatal hypercytokinaemia (with release of IL-1, IL-6, IL-8 and IL-29 with IL-6 being of paramount importance), with multi organ failure characterized by ARDS (Acute respiratory distress syndrome from lung inflammation), in some cases myocarditis and very rarely acute necrotizing encephalopathy.

- Leronlimab is generating some interest and according to Bruce Patterson, CEO of IncelDX , severely ill Covid patients show high levels of RANTES and Leronlimab shows a considerable decrease in plasma viral load. https://www.youtube.com/watch?v=YAZiy2CTqZo&list=PLu_dbA SO_frA2LbNeqX-FR88dHa9zWgLI&index=4&t=332s
Leronlimab is a CCR5 antagonist and was originally developed to treat HIV. CCR5 is a chemokine receptor found on the surface of monocytes and lymphocytes. CCL5, also known as RANTES, is the ligand for CCR5 receptor. RANTES is a chemotactic cytokine which is a chemoattractant for T cells, Eosinophils and Basophils and plays an active role in recruiting leucocytes into inflammatory sites. According to a manuscript published by Dr. Bruce Patterson,https://www.medrxiv.org/content/10.1101/2020.05.0 2.20084673v1.full.pdf (Disruption of CCL5/RANTES – CCR5 pathway), critically ill Covid patients have a profound elevation of plasma IL-6, IL-1β, IL-8 and CCL5 with levels of CCL5 showing correlation with acute renal failure and liver toxicity. The paper posited that SARS-Cov-2 infected airway epithelial cells and macrophages expressed high levels of CCL5 and blocking the CCR5-CCL5/RANTES pathway would hinder the migration of pro-inflammatory leucocytes thereby mitigating the pulmonary hyperinflammation.
- STI-1499 – Researchers at Sorrento Therapeutics have isolated a new antibody dubbed STI-1499 that completely inhibits SARS-

Cov-2 virus in cell culture. They were looking for antibodies that block the interaction of S1 protein (S1 domain of the Spike protein) with ACE2 receptor which is responsible for viral entry into cells. Among the antibodies showing neutralizing activity, one antibody stood out for its ability to completely block viral entry into healthy cells – STI1499. Moreover it seemed to be able to completely neutralize the virus at a very low dose and is being explored as a potential stand-alone therapy owing to its high potency (most other antibodies are being tested for their efficacy against the SARS-Cov-2 virus in conjunction with small-molecule drugs or fusion proteins.

Considering the equivocal outcome of clinical trials with approximately 47 drugs including those which initially seemed to hold promise for the treatment of Covid-19 like Remdesivir (Remdesivir has recently been approved by the FDA although its effectiveness is a contentious issue), Hydroxychloroquine, Azithromycin, Lopinavir/Ritonavir, Oseltamivir, Ivermectin etc., and the unavoidable delay in Covid-19 vaccine development despite frantic efforts through "Operation Warp speed," serious effort is being focused on exploring the possibility of mAbs directed against the viral immunogenic epitopes as a feasible therapeutic intervention. "I think monoclonal antibody therapy has enormous promise as the next big thing for Covid-19," says Dr. Peter Hotez of Bayor University School of Medicine. If monoclonal antibodies can be developed before a vaccine, this would be an advantage in the current situation where time is of the essence. Regeneron pharmaceuticals is hoping to soon start clinical trials with monoclonal antibodies for the treatment of Covid-19 in humans. For this they are using a novel method of monoclonal antibody production which involves the use of convalescent human plasma. (In conventional convalescent plasma therapy, the plasma is directly infused into the patient to provide short-term passive immunity.) For the development of mAbs, researchers screen the convalescent plasma to cull out the highest affinity antibodies. Proteomics and proteogenomics (the details of which are beyond the scope of this book) are used to identify the antigen-binding sequences of these promising antibodies and are correlated with originating nucleotide sequences in reference databases. They can then be cloned to generate monoclonal antibodies against SARS-Cov-2.

It might be possible to even develop nanobodies against the virus, which due to their smaller size can be aerosolized for delivery directly to the lungs. Pulmonary administration could potentially prove to be of great advantage in the treatment of Covid related viral pneumonia and ARDS (acute respiratory distress syndrome). This route of administration (pulmonary/inhaled) is being tested with nebulized anti-IL13 Fab' fragments in murine models of asthma (IL-13 is posited to play a key role in the cascade of events leading to airway inflammation in Bronchial asthma) and with Cetuximab (chimeric anti-EGFR mAb) for lung cancer.

Monoclonal antibody Biosimilars
https://fortune.com/2015/02/06/biosimilars-what-are-they/

Biosimilars are generic versions of the original innovator biological product. Since mAbs are biologics (biologics or biologic drugs are drugs produced from living organisms or components of living organisms), generic versions of mAbs (produced after the patent for the original mAb produced by the innovator company expires) are known as Biosimilar mAbs. They are a new and expanding part of the pharmaceutical industry. Producing generic small molecule drugs is relatively simple compared to the manufacture of biosimilars. Biologics unlike small molecule drugs are complex molecules produced in living cells and their manufacture is a much more painstaking process.

Biosimilars are less costly imitations of biologics. They are used to treat a range of diseases including cancer, rheumatoid arthritis, diabetes, and anemia. But they are different from generic small molecule drugs and their innovator counterparts, in that generic biosimilar drugs are not exact copies of the original innovator biologic.

Biologic drugs are made using living cells that treat disease, usually by genetically modifying cells. They are big and very complex molecules, often 200 to 1,000 times the size of more common small-molecule drugs. For example, aspirin, part of a common category of medicine known as small-molecule drugs, is made of up only 21 atoms. While the biologic drug Enbrel, which is used to treat rheumatoid arthritis and plaque psoriasis, consists of more than 20,000 atoms.

As a result of their complex makeup, biologics are highly sensitive to manufacturing and handling conditions, and many of those production details are highly guarded intellectual property of the company that develops the initial drug. Creating imitations is, therefore, very difficult.

Producing generic small-molecule drugs is relatively simple once you know the molecular structure of the small-molecule drug. The production of Biosimilars is a much more challenging process because living cells are highly sensitive to their environments, and manufacturers have to create their own, unique process to coax these cells to produce an identical outcome to an existing treatment.

To give an analogy, Biosimilars could be compared to snowflakes. The molecular makeup of each biosimila will look unique, like individual snowflakes, even though they all have similar outcomes. This is the result of differing manufacturing processes.

That makes drug approvals challenging. Generics are approved based on matching chemical structure, but that doesn't work for biosimilars. Each new biosimilar has to undergo clinical trials (as opposed to generic small molecule drugs which only have to undergo Bioavailability and Bioequivalence studies) to prove the outcome matches that of the biologic its imitating even though it looks structurally different, according to recently announced guidelines from the Food and Drug Administration.

Much like generics, biosimilars can help cut drug costs, though the savings are smaller because of their complexity as well as regulatory challenges of getting FDA approvals. Biosimilars cost about $75 million to $250 million to reach the approval stage, versus around $2 million to $3 million for a generic small-molecule medicine, according to biologics-maker Sandoz, a Novartis company.

The naming system for monoclonal antibody biosimilars will follow FDA guidelines according to which the name of the original mAb (the generic or core name) will be followed by a four-letter suffix indicating the name of the manufacturer. E.g. The drug marketed under the trade name of Hyrimoz (Adalimumab-adaz) by Sandoz is a mAb biosimilar to the original product Humira (Adalimumab) brought to the market by its innovator company, Abbvie.

> https://fortune.com/2015/02/06/biosimilars-what-are-they/

The "Purple Book" lists biological products, including any biosimilar and interchangeable biological products, licensed by FDA under the Public Health Service Act.

Pharmacokinetics and routes of administration

It is generally known that proteins and mAbs are prone to undergo enzymatic degradation and unfolding especially in the GIT. Oral delivery systems for these biologically active compounds are hence challenging to develop, unlike most small molecule drugs. Thus, the routes of administration of biologics are normally parenteral injection (intravenous, subcutaneous, intramuscular or intradermal). Most proteins and mAbs are now formulated for subcutaneous injection

Some are administered intravenously (e.g., infliximab), some can be administered subcutaneously (e.g., emicizumab), and some can be administered by either route (e.g., rituximab, in different formulations). Intramuscular use has also been reported (e.g., palivizumab). Subcutaneous administration is a desired route of administration Antibodies injected subcutaneously are taken up by lymphatic channels and may not reach maximum plasma concentration for several days.

mAbs are limited in their ability to penetrate and accumulate in tissues due to their large size and are therefore likely to be restricted to the interstitial space after injection. In general, biotherapeutics can reach blood circulation by two pathways: via blood capillaries or lymphatic vessels. Absorption through blood capillaries relies on passive transport and is restricted to compounds with a molecular weight cut-off (MWCO) of up to 16 kDa . Hence, most biotherapeutics will not be able to be transported via capillary routes, but rather rely on the lymphatic system where the protein reaches the systemic circulation at the thoracic duct.

Some of the ways in which these proteins reach the peripheral tissues are: - transport through fenestrae pores on capillary walls, through transcellular pathways via endothelial cells, cellular uptake from surrounding fluid by pinocytosis, receptor mediated endocytosis and phagocytosis by immune cells.

The distribution of therapeutic proteins is limited to plasma rather than tissue. As a consequence, delivering proteins to maintain effective therapeutic concentration at target tissues is challenging. For example, bevacizumab and ranibizumab are given intravitreally. Owing to the size of both anti-VEGF medicines, it is unlikely to deliver a significant level of both drugs to the posterior segment of the eye by systemic administration, where the blood retina barrier (BRB) is a major barrier for drug transport. The subcutaneous administration of trastuzumab is enabled by the use of recombinant human hyaluronidase (rHuPH20), which behaves as a permeation enhancer.

The use of antibody fragments in lieu of full- length antibodies would enhance the efficiency of penetration into tissues like tumor masses.

The elimination of mAb therapeutics can be accelerated by immunogenicity and the development of antibody-drug antibodies (ADAs), particularly with those proteins derived from animals. ADAs can also cause acute hypersensitivity or infusion reactions. Furthermore, ADAs can also competitively bind to the active region of the therapeutic protein such as the receptor-binding site to neutralize the antibody drug, therefore compromising efficacy. ADAs can also unpredictably change drug pharmacokinetic properties, biological effects and the toxicity profile. Humanized and fully human mAbs and other therapeutic proteins are less immunogenic in humans compared with non-human derived proteins (e.g., murine antibodies), although humanized proteins are not devoid of immunogenicity.

Certain formulation strategies can be utilized to prolong the duration of monoclonal antibodies such a Hydrogels, Liposomes, Micro/Nanoparticles and Micelles. For detailed information on these drug delivery systems, which is beyond the scope of this book, check

out "An overview of antibody drug delivery"
https://www.ncbi.nlm.nih.gov/pmc/articles/PMC6161251/

Adverse Drug Reactions

mAbs have revolutionized the treatment of cancers and several other diseases. Being a more targeted form of therapy designed to bind specifically to a particular protein involved in the pathogenesis of the disease, they are associated with less adverse effects than most other forms of therapy especially conventional chemotherapy and are generally better tolerated. Nevertheless, they are associated with a wide spectrum of adverse effects, necessitating efforts to identify, describe and manage these reactions in order to ensure their safe use. Being non-endogenous proteins of sufficient size, immunogenicity is always a safety concern and despite progressive efforts at replacing xenogeneic moieties with human components leading to the development of chimeric, humanized and fully human mAbs, there is still potential for developing anti-idiotypic antibodies (anti-idiotype antibodies are antibodies that bind to the variable region of other antibodies) to the SDR regions (only about 20-30% of amino acids in the CDR actually make direct contact with the antigen and these residues are known as Specificity determining Residues or SDRs). Generally, the more humanized the mAb, the less would be its immunogenic potential. E.g. the Trifunctional Bispecific rat/mouse hybrid antibody, Catumaxomab would be more immunogenic than the fully humanized Erenumab. Exceptions exist, e.g. the fully human mAb Daratumumab has an astounding 48% infusion reaction (although infusion reactions are not true allergic reactions).

Common adverse reactions to mAbs are true hypersensitivities (Types I, II, III and IV) ; Anaphylactoid reactions like Infusion reactions (IRs)/

Cytokine Release syndrome (CRS); Tumor Lysis Syndrome (TLS); Pulmonary and Cardiac adverse events; Progressive multifocal leukoencephalopathy and some targeted AEs.

Type I hypersensitivities mediated by IgE antibodies and responsible for anaphylactic reactions characterized by bronchospasm, cardiovascular collapse etc. are relatively uncommon upon administration of mAbs.

Type II hypersensitivities – This is mediated by antibody directed towards antigens present on the surface of cells such as erythrocytes, leucocytes and platelets and probably hematopoietic precursor cells in the bone marrow. This is caused by alteration of molecules on the cell surface by the drug molecule (which acts as a hapten) thus inducing an autoantibody response. Cases of thrombocytopenia, mAb induced late onset neutropenia (LON) and Autoimmune hemolytic anemia (AIHA) have been observed following treatment with mAbs. Sometimes it may be difficult to distinguish if the reaction is a Type II hypersensitivity to the mAb or is attributable to another etiology. For e.g. mAbs are frequently given in combination with conventional chemotherapeutic agents which have a well-known adverse effect of suppression of megakaryopoiesis, agranulocytosis etc. It could also be part of the natural progression of a lymphoproliferative disease. It could also be due to the pharmacologic mechanism of action of the drug (mAbs that target CD20 may cause thrombocytopenia due to the presence of CD20 on platelets).

Type III hypersensitivity – Also called Serum sickness, it is due to the deposition of Ag-Ab complexes (immune complexes) in tissues giving rise to symptoms such as fever, arthralgia, lymphadenopathy etc. Chimeric mAbs have the potential to induce serum sickness like reaction due to their higher content of murine residues and consequent greater immunogenicity.

Type IV hypersensitivity – This is mediated by T-cells unlike the other hypersensitivity reactions which are primarily mediated by a humoral response. Delayed type hypersensitivity reactions to mAbs include cutaneous reactions such as maculopapular exanthema, Fixed drug eruptions (FDE) – erythematous patches following systemic drug administration that recur at the same site, Erythema multiforme (EM) – with typical target lesions showing erythematous ring around pale center, Steven Johnson syndrome (SJS) etc.

Anaphylactoid reactions which are not immune – mediated but are due to release of bradykinin etc. They are called Infusion reactions (IRs) or Cytokine Release Syndromes (CRS) characterized by flu-like symptoms within a few hours of infusion. It is caused by the release of large amounts of cytokines causing massive systemic inflammation leading sometimes to ARDS and multi organ failure and sometimes death.

Tumor Lysis Syndrome (TLS) – Associated with any anti-cancer therapy due to the rapid death of large numbers of malignant cells. It usually occurs 48 – 72 hrs after the start of anti-neoplastic treatment. Lysis of large numbers of cells lead to profound ionic imbalance (hyperkalemia, hypercalcemia, hyperphosphatemia, hyperuricemia) which can cause cardiac arrhythmias and seizures.

Adverse reactions related to pharmacologic target molecule – mAbs that bind to EGFR can cause mucocutaneous reactions, mAbs that bind to VEGF can cause thromboembolic events and those that target the CD20 antigen can thrombocytopenia since the CD20 antigen is also found on platelets.

Cardiac adverse events - arrhythmias, cardiomyopathy and LVD (left ventricular dysfunction) have been known to occur as mAb AEs.

Pulmonary adverse events – These comprise a heterogenous group of lung diseases often classified under the title of Drug Induced Lung Diseases (DILD) including pneumonitis, Bronchiolitis Obliterans Organizing Pneumonia (BOOP). A case of fatal intra-alveolar hemorrhage has been reported with Rituximab administration.

Autoimmune diseases – Autoimmune diseases caused by mAbs are rare. Ipilimumab has been shown to cause autoimmune enterocolitis. There has been no mention of drug-induced lupus in current literature.

Progressive Multifocal Leukoencephalopathy (PML) – This is a progressive, usually fatal viral disease that in some respects resemble multiple sclerosis, as the myelin sheath of nerve cells is ultimately destroyed affecting transmission of nerve impulses. It is caused by the JC virus or Human Polyoma virus 2 which persists asymptomatically in more than one-third of the population. JC virus induced demyelination occurs in immunosuppressed states like transplant recipients on immunosuppressants, HIV patients etc. It is also occasionally seen upon administration of mAbs whose mechanism of action entails the destruction of B cells.

Evaluation and mitigation of AEs – Skin prick and intradermal testing can often be used to detect immediate Type I and delayed Type IV reactions to drugs. True Type I reactions are mediated by IgE antibodies and immunoassays to detect IgE specific for individual mAbs can be developed. Patch testing is both a screening test for hypersensitivity and a provocation test in the skin and can be used to investigate delayed cutaneous reactions. In-vitro tests that detect platelet-reactive serum antibodies are available to aid the diagnosis of drug-induced thrombocytopenia. Anti-neutrophil antibody tests are used to help in the diagnosis of immune-mediated neutropenia and agranulocytosis. IgG/IgM antibodies to drug-cell membrane complex can be detected in cases of drug-induced anemia. Drug induced lung diseases can be evaluated using HR CT, Pulmonary function testing and Bronchoscopy with BAL (Bronchoalveolar lavage). ELISPOT cytokine assays (ELISA tests for detecting cytokines) are useful for evaluating T lymphocyte associated mucocutaneous reactions and detecting drug-reactive T cells.

Further research is needed to develop new and improved tests for the accurate diagnosis of adverse reactions and elucidation of their mechanism of action.

The Theralizumab disaster – TG1412 (Theralizumab) was developed as a super agonist to the CD28 protein on the surface of the regulatory T cells. It can bind to CD28 on the regulatory T cells thereby activating them (Activation of regulatory T cells play a useful role in a variety of autoimmune disorders and cancer). In pre-clinical studies (done in non-human primates) the drug had demonstrated some therapeutic potential and seemed well- tolerated. However, in its FIH (First in human) trials, where 1/500th of the dose was used based on NOAEL (No Observed Adverse Effect Level) and MABEL (Minimal Anticipated Biological Effect Level), within minutes of infusion, all six healthy volunteers developed severe symptoms of Cytokine Release Syndrome culminating in multiorgan failure.
https://www.youtube.com/watch?v=a9_sX93RHOk (This is a documentary of the Theralizumab disaster).

This disaster highlights the fact that Pharmacokinetic (ADME processes) and Pharmacodynamic (MOA) parameters in human volunteers cannot be extrapolated from animal models in all cases.

One of the main factors in immunogenicity of mAbs, apart from the antigenicity of any foreign protein in general is the route of administration. There is evidence that intravenous administration has a lower risk of eliciting immune reactions compared to intramuscular and subcutaneous routes. It has also been reported that higher drug doses are less likely to provoke an immune reaction than lower doses, an effect known as "high zone tolerance". (https://www.jacionline.org/article/S0091-6749(15)00411-X/fulltext}. Immunocompromised patients, of course are less likely to develop Anti-drug antibodies (ADAs) against any foreign protein compared to immunocompetent persons. So, the immune status of the patient is a factor in the formation of ADAs. The clinical application of the mAb is also a factor in determining its potential for evoking ADAs. mAbs directed against immune cells (e.g. Rituximab used to treat NHL destroys healthy immune cells in addition to malignant immune cells and is therefore less likely to induce the production of ADAs.

Tocilizumab, a DMARD for RA, currently under trials for treatment for COVID-19 pneumonia, shows clear evidence of higher risk of heart attacks, strokes and heart failure compared to other DMARDs, according to STAT news which advocates changes to the drug's label to warn the public.

Business Perspective

According to the Journal of Pharmaceutical Innovation, until 2014, 47 monoclonal antibodies had been approved in the US and EU. From the year 2008 – 2013, the market grew from approximately $39 Billion to approximately $75 Billion. Extrapolating from this growth rate, it is safe to conjecture that the sales of monoclonal antibodies are likely to reach $125 Billion by 2020 and $138.6 Billion by 2024. The two major fields of medicine where monoclonal antibodies have shown unparalleled promise are Oncology and Autoimmune diseases.

According to "Creative Biolabs", the major target areas for monoclonal antibody therapy are:-

- Oncology (including hematological cancer and non-hematological cancer) – 34%
- Immune disease mainly autoimmune disease – 26%
- Infectious diseases – 4%
- Cardiovascular disease – 3%
- Orthopedic diseases, eye diseases and rare diseases – 6%

According to "Market Data Forecast", companies leading the global monoclonal antibody market are :-

1. GlaxoSmithKline
2. Novartis
3. Pfizer
4. Thermo Fischer Scientific
5. Eli Lilly
6. Seattle Genetics
7. Bristol-Myers Squibb
8. La Roche Pharmaceuticals
9. Teva Pharmaceuticals
10. Shangai Junshi Biosciences

According to the article "Antibodies to watch in 2019", there are over 570 antibody drugs in various phases of clinical development (from pre-clinical studies to Phase III) with 62 of them in late stage clinical development (Phases II and III).

To obtain further detailed information on Biopharmaceutical products including monoclonal antibodies, BioPharma is a good source
http://www.biopharma.com/cgi/results7.lasso

Top Ten Monoclonal antibodies by sales in 2018
https://biopharmadealmakers.nature.com/users/9880-biopharma-dealmakers/posts/53687-moving-up-with-the-monoclonals

- Humira – Adalimumab
- Keytruda – Pembrolizumab
- Opdivo – Nivolumab
- Stelara – Ustekinumab
- Ocrevus – Ocrelizumab
- Cosentyx – Secukinumab
- Dupixent – Dupilumab
- Darzalex – Daratumumab
- Perjeta – Pertuzumab
- Entyvio - Vedolizumab

References: -

- Making sense of monoclonal antibodies – Pharmacy Times
  https://www.pharmacytimes.com/contributor/brandon-dyson-pharmd-bcps/2016/08/making-sense-of-monoclonal-antibodies
- Monoclonal antibodies to treat cancer – American Cancer Society
  https://www.cancer.org/treatment/treatments-and-side-effects/treatment-types/immunotherapy/monoclonal-antibodies.html
- Transgenic animal – an overview : Science Direct : From the Journal of Basic and Applied Bone biology

- https://www.sciencedirect.com/topics/medicine-and-dentistry/transgenic-animal
- Transfer gene – Stem cell biology and Tissue engineering in Dental sciences https://www.sciencedirect.com/topics/biochemistry-genetics-and-molecular-biology/transfer-gene
- The history of monoclonal antibody development – Progress, remaining challenges and future innovations: Annals of Medicine and Surgery https://www.sciencedirect.com/science/article/pii/S2049080114000624
- Microinjection – methods in cell biology https://www.sciencedirect.com/topics/medicine-and-dentistry/microinjection
- List and types of monoclonal antibodies (FDA approved): Medicinenet.com https://www.medicinenet.com/monoclonal_antibodies/article.htm#what_are_human_monoclonal_antibodies?
- Biosimilars – what are they? : Fortune magazine https://fortune.com/2015/02/06/biosimilars-what-are-they/
- List of therapeutic monoclonal antibodies – Wikipedia https://en.wikipedia.org/wiki/List_of_therapeutic_monoclonal_antibodies
- Hybridoma production – EuroMabNet https://www.euromabnet.com/protocols/hybridoma.php
- Epstein Barr virus hybridoma technique – SpringerLink https://link.springer.com/chapter/10.1007/978-1-4684-4949-5_4
- Human-human hybridomas and human monoclonal antibodies obtained by fusion of lymph node lymphocytes from breast cancer patients: Cancer Research

https://cancerres.aacrjournals.org/content/canres/48/11/3208.full.pdf
- Production of human hybridomas secreting antibodies to measles virus – Nature (International journal of science) https://www.nature.com/articles/288488a0
- Hybridoma vs Phage display for monoclonal antibody production – Technologynetworks.com https://www.technologynetworks.com/drug-discovery/blog/hybridoma-vs-phage-display-for-monoclonal-antibody-production-which-technique-for-which-purpose-314741
- Phage display technology for the production of recombinant monoclonal antibodies – Labome https://www.labome.com/method/Phage-Display-Technology-for-the-Production-of-Recombinant-Monoclonal-Antibodies.html
- Production of recombinant antibodies using bacteriophage – European journal of Microbiology and Immunology https://www.ncbi.nlm.nih.gov/pmc/articles/PMC4029287/
- Selection and identification of mimic epitopes for gastric cancer associated antigen-MG7Ag – Molecular Cancer Therapeutics https://pubmed.ncbi.nlm.nih.gov/12657725/
- Mimic epitope recognized by mAb MG7 against gastric cancer through screening phage displayed random peptide library – Chinese journal in PubMed https://pubmed.ncbi.nlm.nih.gov/11798777/
- Fundamental technologies for antibody production-Phage Display: Creative Biolabs https://www.creative-biolabs.com/blog/index.php/fundamental-technologies-for-antibody-production-phage-display/
- DNA libraries and generating cDNA – Khan Academy https://www.khanacademy.org/test-

- prep/mcat/biomolecules/dna-technology/v/dna-libraries-generating-cdna
- Molecular insights into fully human and humanized monoclonal antibodies – The journal of Clinical and Aesthetic Dermatology
  https://www.ncbi.nlm.nih.gov/pmc/articles/PMC5022998/
- Hybridoma – an overview: ScienceDirect
  https://www.sciencedirect.com/topics/medicine-and-dentistry/hybridoma
- Enbrel-mechanism of action: Enbrel.com
  https://www.enbrelpro.com/more-about-enbrel/mechanism-of-action
- Mouse myeloma proteins – Millipore sigma
  https://www.sigmaaldrich.com/life-science/cell-biology/antibodies/antibody-products.html?TablePage=9674682
- Diabody – SpringerLink
  https://link.springer.com/referenceworkentry/10.1007%2F978-3-642-16483-5_1603 Ozaki S. (2011) Diabody. In: Schwab M. (eds) Encyclopedia of Cancer. Springer, Berlin, Heidelberg
- Chimeric antibodies – SpringerLink , Encyclopedia of cancer
  https://link.springer.com/referenceworkentry/10.1007%2F978-3-642-16483-5_1091
- Humanized antibodies – SpringerLink, Encyclopedia of cancer
  https://link.springer.com/referenceworkentry/10.1007%2F978-3-642-16483-5_2863
- Human monoclonal antibody-an overview/ScienceDirect
  https://www.sciencedirect.com/topics/medicine-and-dentistry/human-monoclonal-antibody
- Optimized expression of full-length IgG_1 antibodies in a common E.Coli strain PLOS/ONE

https://journals.plos.org/plosone/article?id=10.1371/journal.pone.0010261
- Recombinant IgG expression in mammalian cells https://pdfs.semanticscholar.org/0257/bdd9a6821b27aab13c483a23ed7a4e439732.pdf
- Difference between a cloning vector and an expression vector – studiousguy.com https://studiousguy.com/difference-between-a-cloning-vector-and-an-expression-vector/
- Transformation (genetics) – Wikipedia https://en.m.wikipedia.org/wiki/Transformation_(genetics)
- Choosing the right E.Coli strain for transformation – New England Biolabs Inc. https://bitesizebio.com/30630/choosing-right-e-coli-strain/
- How DNA is inserted with microinjections – The Balance https://www.thoughtco.com/microinjection-375568
- Microbials for the production of monoclonal antibodies and antibody fragments – Trends in Biotechnology https://www.ncbi.nlm.nih.gov/pmc/articles/PMC3906537/
- DNA Synthesis - How products are made http://www.madehow.com/Volume-6/DNA-Synthesis.html
- Monoclonal antibodies – a review: Bentham Science http://www.eurekaselect.com/154790/article
- Cell lines – Thermo Fischer scientific https://www.thermofisher.com/us/en/home/references/gibco-cell-culture-basics/cell-lines.html
- Hybridoma fusion partners cell lines – Millipore sigma https://www.sigmaaldrich.com/technical-documents/protocols/biology/cell-culture/hybridoma-fusion.html
- GenBank Overview – www.ncbi.nlm.gov
- Gene machine, general biotechnology – Biocyclopedia

https://biocyclopedia.com/index/biotechnology/genes_genetic_engineering/genes_nature_concept_and_synthesis/biotech_gene_machine.php
- Antibody phage display, technique and applications – Journal of Investigative Dermatology
  https://www.ncbi.nlm.nih.gov/pmc/articles/PMC3951127/
- Overview of therapeutic monoclonal antibodies – UpToDate
  https://www.uptodate.com/contents/overview-of-therapeutic-monoclonal-antibodies/print
- Antibody drug nomenclature – Bioatla
  https://www.bioatla.com/appendix/antibody-nomenclature/
- Monoclonal antibodies, a business perspective – mAbs (journal name)
  https://www.ncbi.nlm.nih.gov/pmc/articles/PMC2725420/
- Monoclonal antibody therapy – Wikipedia
  https://en.wikipedia.org/wiki/Monoclonal_antibody_therapy
- Monoclonal antibody market 2019-2025, growth, key players, size, demands and forecasts – Reuters Plus
  https://www.reuters.com/brandfeatures/venture-capital/article?id=101636
- The therapeutic monoclonal antibody market – mAbs
  https://www.ncbi.nlm.nih.gov/pmc/articles/PMC4622599/
- Antibody-directed enzyme prodrug therapy, efficiency and mechanism of action in colorectal carcinoma – AACR (American Association for Cancer Research)
  https://clincancerres.aacrjournals.org/content/6/3/765
- Anti-HER2 Immunoliposomes, enhanced efficacy attributable to targeted delivery – AACR
  https://clincancerres.aacrjournals.org/content/8/4/1172

- Definition of immune checkpoint inhibitor – NCI (National Cancer Institute) dictionary of cancer terms: NIH (National institute of health) https://www.cancer.gov/publications/dictionaries/cancer-terms/def/immune-checkpoint-inhibitor
- New monoclonal antibody approvals hit record levels in 2017 – Creative Biolabs https://www.creative-biolabs.com/blog/index.php/new-monoclonal-antibody-drug-approvals-in-2017/
- Adverse events to monoclonal antibodies used for cancer therapy – Oncoimmunology https://www.ncbi.nlm.nih.gov/pmc/articles/PMC3827071/
- Mechanisms of drug induced allergy – Mayo Clinic Proceedings https://www.ncbi.nlm.nih.gov/pmc/articles/PMC2664605/
- Cytokine detection by ELISPOT – Diabetes/Metabolism research and reviews https://pubmed.ncbi.nlm.nih.gov/12397579/
- The immunogenicity of humanized and fully human antibodies: - mAbs https://www.ncbi.nlm.nih.gov/pmc/articles/PMC2881252/
- Frontiers in immunology https://www.frontiersin.org/articles/10.3389/fimmu.2019.02290/full
- Ablynx – A Sanofi company https://www.ablynx.com/our-company/overview/
- Amgenscience.com https://www.amgenscience.com/
- Plos.org https://plos.org/
- Labome.com https://www.labome.com/index.html
- Clinical and experimental immunology

- https://www.researchgate.net/journal/1365-2249_Clinical_Experimental_Immunology
- Encyclopedia of genetics – Schildkraut
  https://www.sciencedirect.com/topics/neuroscience/genetic-engineering
- Introduction of biotechnology – W.T Godbey
  https://bagitds.files.wordpress.com/2017/01/agr-203_an_introduction_to_biotechnology.pdf
- G – Biosciences
  https://www.gbiosciences.com/
- Trends in Biotechnology – Elseiver
- Recent advances in Monoclonal antibody therapies for multiple sclerosis – Expert Opinion on Biological Therapy
  https://www.ncbi.nlm.nih.gov/pmc/articles/PMC4913471/
- Natalizumab: A new treatment for relapsing remitting multiple sclerosis – Therapeutics and Clinical risk management
  https://www.ncbi.nlm.nih.gov/pmc/articles/PMC1936307/
- Moving up with the monoclonals – Biopharma dealmakers
  https://biopharmadealmakers.nature.com/users/9880-biopharma-dealmakers/posts/53687-moving-up-with-the-monoclonals

- The Effect of Intravitreal Injection of Bevacizumab on Retinal Circulation in Patients with Neovascular Macular Degeneration – IOVS https://iovs.arvojournals.org/article.aspx?articleid=2165611
- Overview of therapeutic monoclonal antibodies – Uptodate
  https://www.uptodate.com/contents/overview-of-therapeutic-monoclonal-antibodies/print
- Antibody therapeutics approved or in regulatory review in the EU or US – Welcome to the antibody society

https://www.antibodysociety.org/resources/approved-antibodies/ (an excel file can be downloaded at this site for all therapeutic Mabs approved or under review in the US or EU).
- Soluble Angiotensin-Converting Enzyme 2: A Potential Approach for Coronavirus Infection Therapy? – Clinical Science https://pubmed.ncbi.nlm.nih.gov/32167153/
- Overview of Antibody Drug Delivery – Pharmaceutics https://www.ncbi.nlm.nih.gov/pmc/articles/PMC6161251/
- Creative Biolabs https://www.creative-biolabs.com/blog/index.php/new-monoclonal-antibody-drug-approvals-in-2017/
- Biopharma http://www.biopharma.com/cgi/results7.lasso
- Neutralizing Monoclonal Antibodies to Fight HIV-1: On the Threshold of Success – Frontiers in Immunology https://www.frontiersin.org/articles/10.3389/fimmu.2016.00661/full
- Economic Analysis of Batch and Continuous Biopharmaceutical Antibody Production: A Review – Journal of Pharmaceutical Innovation https://www.ncbi.nlm.nih.gov/pmc/articles/PMC6432653/
- Monoclonal Antibodies Market – Market Data Forecast https://www.marketdataforecast.com/market-reports/global-monoclonal-antibodies-market
- Antibodies to watch in 2019 https://www.tandfonline.com/doi/full/10.1080/19420862.2018.1556465
- TGN1412: From Discovery to Disaster – Journal of young pharmacists https://www.ncbi.nlm.nih.gov/pmc/articles/PMC2964774/
- Wuhan Data Link COVID-19 With Myocardial Damage

https://www.medscape.com/viewarticle/927636?nlid=134701_32 43&src=WNL_mdplsfeat_200331_mscpedit_imed&uac=297939CR &spon=18&impID=2330911&faf=1
- Woman with COVID-19 developed a rare brain condition. Doctors suspect a link.
  https://www.livescience.com/woman-with-covid19-coronavirus-had-rare-brain-disease.html
- COVID-19: consider cytokine storm syndromes and immunosuppression
  https://www.thelancet.com/journals/lancet/article/PIIS0140-6736(20)30628-0/fulltext
- Genentech Launches Phase III Trial of Actemra as Coronavirus Treatment
  https://www.genengnews.com/virology/coronavirus/genentech-launches-phase-iii-trial-of-actemra-as-coronavirus-treatment/
- Kevzara (sarilumab)
  https://www.centerwatch.com/directories/1067-fda-approved-drugs/listing/3702-kevzara-sarilumab
- Failure to warn: Hundreds died while taking an arthritis drug, but nobody alerted patients
  https://www.statnews.com/2017/06/05/actemra-rheumatoid-arthritis-fda/
- Changes to International Nonproprietary Names for antibody therapeutics 2017 and beyond: of mice, men and more – mAbs
  https://www.ncbi.nlm.nih.gov/pmc/articles/PMC5590622/
- Nomenclature of humanized mAbs: Early concepts, current challenges and future perspectives – Human antibodies
  https://www.ncbi.nlm.nih.gov/pmc/articles/PMC6294595/
- serological assay to detect SARS-CoV-2 seroconversion in humans

https://www.medrxiv.org/content/10.1101/2020.03.17.20037713v1.full.pdf
- Baculovirus Expression Vectors
https://link.springer.com/chapter/10.1007/978-1-4899-1834-5_14
- Vector: pCAGGS
https://plasmid.med.harvard.edu/PLASMID/GetVectorDetail.do?vectorid=184
- https://www.livescience.com/testing-old-drugs-covid-19.html?utm_source=Selligent&utm_medium=email&utm_campaign=17470&utm_content=20200504_Coronavirus_Infographic+&utm_term=2594286&m_i=PnkqlXXwqXyyDli0OA0btOcLKN5XA2uirtBhb9NyxiYlStsoxQKp3DKajv%2Bq6tI6WxnqvN66GGGseh2zr1dwnTx1ic7d5y
- https://link.springer.com/article/10.1007/s40259-019-00392-z
- https://www.clinicaltrialsarena.com/news/cytodyn-leronlimab-covid-19-data/
- https://www.globenewswire.com/news-release/2020/05/04/2026582/0/en/FDA-Approves-54-Emergency-INDs-for-Leronlimab-Treatment-of-Coronavirus-CytoDyn-Requests-Compassionate-Use-from-FDA-for-COVID-19-Patients-Not-Eligible-for-Participation-in-Two-Ongoi.html
- https://www.precisionvaccinations.com/leronlimab-therapy-checked-covid-19-inflammation
- https://www.drugtargetreview.com/news/60206/celltrion-selects-14-lead-monoclonal-antibodies-for-covid-19-treatment/
- https://www.antibodysociety.org/covid-19/
- http://www.sci-news.com/medicine/sti-1499-antibody-sars-cov-2-coronavirus-08438.html

- https://science.sciencemag.org/content/early/2020/05/19/science.abc6284
- https://pubmed.ncbi.nlm.nih.gov/22904197/
- https://www.medrxiv.org/content/10.1101/2020.05.02.20084673v1.full.pdf

Updated May 30th

Made in the USA
Coppell, TX
29 August 2021